NORTHFIELD BRANCH
847-446-5990

NATURAL DISASTER!

All About Heat Waves and Droughts

Discovering Earth's Scorching Weather

Steve Tomecek

Children's Press®
An imprint of Scholastic Inc.

Content Consultant

Dr. Kristen Rasmussen
Assistant Professor
Department of Atmospheric Science
Colorado State University

Dedication: *In memory of Dr. John Loret and Dr. Walter S. Newman, two visionary scientists who taught me to think about the future of our planet.*

Library of Congress Cataloging-in-Publication Data
Names: Tomecek, Steve, author.
Title: All about heat waves and droughts / by Steve Tomecek.
Description: First edition. | New York : Children's Press, an imprint of Scholastic Inc., 2021. | Series: A true book: natural disaster! | Includes bibliographical references and index. | Audience: Ages 8–10. | Audience: Grades 4–6. | Summary: "This book shows readers the awesome power of heat waves and droughts"— Provided by publisher.
Identifiers: LCCN 2021003964 (print) | LCCN 2021003965 (ebook) | ISBN 9781338769579 (library binding) | ISBN 9781338769586 (paperback) | ISBN 9781338769593 (ebook)
Subjects: LCSH: Heat waves (Meteorology)—Juvenile literature. | Climatic changes—Juvenile literature. | Droughts—Juvenile literature.
Classification: LCC QC981.8.A5 T66 2021 (print) | LCC QC981.8.A5 (ebook) | DDC 551.5/253—dc23
LC record available at https://lccn.loc.gov/2021003964
LC ebook record available at https://lccn.loc.gov/2021003965

10 9 8 7 6 5 4 3 2 1   22 23 24 25 26

Printed in the U.S.A. 113
First edition, 2022

Series produced by Priyanka Lamichhane
Book design by Kathleen Petelinsek
Illustrations on pages 42–43 by Gary LaCoste

Front cover: Background: Extreme heat and drought scorches the landscape, causing cracks in the dry ground; top: Drought takes hold in an area of the Amazon rain forest; top right: A child watches an egg on a sidewalk to see if it will cook on hot pavement during a heat wave; bottom: Wildfires burn trees as firefighters try to contain them.

Back cover: Extreme heat causes the road to crack on a highway.

Find the Truth!

Everything you are about to read is true *except* for one of the sentences on this page.

Which one is **TRUE**?

T or F   Sometimes human activity can cause droughts to happen.

T or F   Heat waves only happen in the summer.

Find the answers in this book.

What's in This Book?

Trees have rings that give clues about when droughts happened in the past.

It is important to stay cool when it's very hot outside.

Droughts in the Rain Forest

A rain gauge measures rainfall.

This farm in Australia has been destroyed by drought.

Weather to the Extreme

In January 2019, Wayne Dunford didn't think things could get much worse. For almost two years, Australia had been going through one of the worst **droughts** in history. Wayne was a farmer. The soil was so dry that he could no longer grow crops and most of the grass had died. Wayne had to feed his animals by hand to keep them from starving. Then, it happened. The country was hit with an extreme **heat wave**. Temperatures

Farmer Wayne Dunford pulls bales of hay to feed his cattle on his farm in Australia.

soared past 115 degrees Fahrenheit (46 degrees Celsius). By the time it was over, millions of animals had died, and **wildfires** had burned thousands of square miles of land. Dunford's fields were completely dry. Because of **global warming**, heat waves and droughts could get even worse. This could lead to **disastrous consequences** for plants, animals, and people. Read on to learn what causes heat waves and droughts and what you can do to prepare for them.

Scientists who measure and predict the weather are called **meteorologists.**

This map shows air temperatures across the United States on September 6, 2020, when much of the Southwest roasted in a dramatic heat wave.

Temperature Key
- 41°F (5°C)
- 59°F (15°C)
- 77°F (25°C)
- 95°F (35°C)
- 113°F (45°C)

The Wonders of Weather

When scientists talk about weather, they are talking about the conditions in the air at a certain place. Two of the conditions that scientists measure are the temperature of the air and the amount of **precipitation**—rain, snow, or hail—that falls from the sky. A heat wave is a stretch of time when the temperature is much higher than normal. A drought is a stretch of time when there is very little precipitation.

Rain forest

Hot desert

Cold desert

A rain forest has a hot, wet climate. In both hot and cold deserts, it rarely rains, so these areas are dry.

Climate

Scientists use the word "**climate**" to describe what the weather is usually like in one area. Places that have a lot of rain have **humid** climates. Places that have hardly any rain have dry climates. Still other places have cold climates that only warm up a little during the summer. Heat waves and droughts happen when the temperature and precipitation are hotter and drier than what's normal for an area.

What Makes the Air Warm?

The air above Earth is called the **atmosphere**. Our atmosphere is heated by Earth's surface. When sunlight hits the surface, it heats up the land and water. The heat from the warm surface rises. Some of this heat is trapped by gases in the atmosphere, like a blanket, and keeps the heat from going out into space. This is called the **greenhouse effect**.

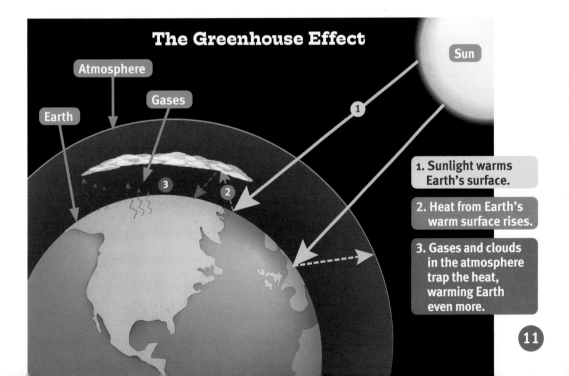

The Greenhouse Effect

Sun

Atmosphere

Gases

Earth

1. Sunlight warms Earth's surface.

2. Heat from Earth's warm surface rises.

3. Gases and clouds in the atmosphere trap the heat, warming Earth even more.

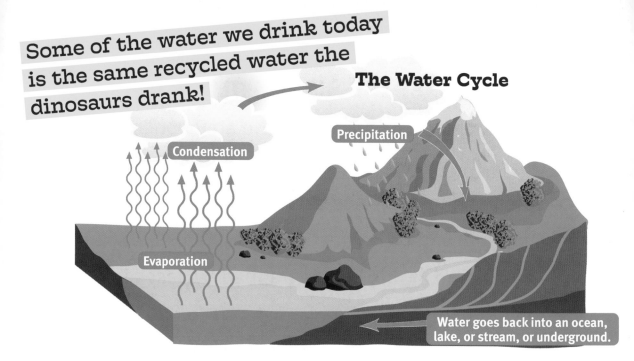

Some of the water we drink today is the same recycled water the dinosaurs drank!

The Water Cycle

Condensation

Precipitation

Evaporation

Water goes back into an ocean, lake, or stream, or underground.

Here Comes the Rain!

Water on Earth moves through a process called the water cycle. The water cycle has three main parts: **Evaporation** is when water on the surface becomes a gas in the air called **water vapor**. As water vapor rises, it cools and turns into tiny drops of liquid water, forming clouds. This is called **condensation**. Over time, the water falls back to the surface as precipitation, and the cycle starts again.

Climate Regions on Earth

Scientists divide Earth's surface into different **climate regions** based on their normal weather: the average temperature and the average amount of precipitation each area gets. Some environments, such as tropical regions, are naturally hot, while others, like polar regions, are cold. In the same way, some environments, like deserts, are naturally dry while others, such as rain forests, get lots of rain.

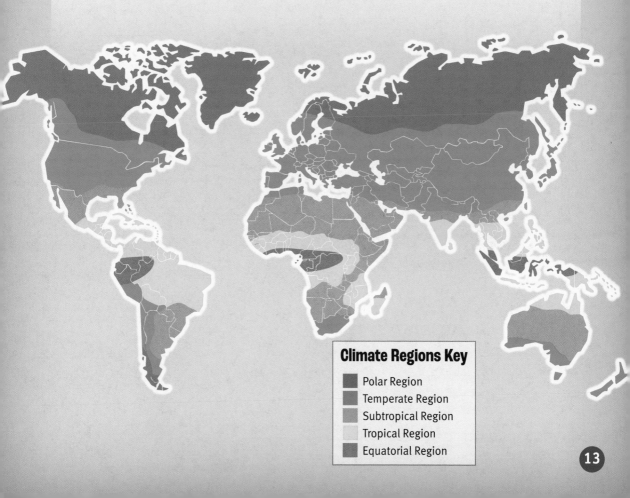

Climate Regions Key

- Polar Region
- Temperate Region
- Subtropical Region
- Tropical Region
- Equatorial Region

More than 2,000 record-high temperatures were set during a heat wave in the central United States in July 2012.

During a heat wave, some days get so hot that you could almost cook an egg on the sidewalk!

The Heat Is On

Heat waves can happen any time of the year, not just in summer. And they don't just happen on land. They can happen in oceans, too. There have even been heat waves in the Arctic! No matter where they happen, heat waves have an impact on people and places.

What Causes Heat Waves?

Earth's atmosphere is made up of large chunks of air called air masses. Sometimes a mass of heavy air, or high-pressure air, gets stuck over an area. Sinking air within the high-pressure area warms, and the hot ground causes the air to get warmer. This causes a heat wave. It lasts until the heavy air eventually moves. New, cooler air comes in, ending the heat wave.

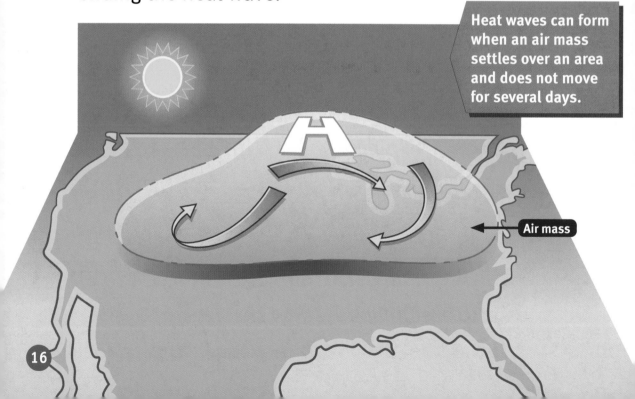

Heat waves can form when an air mass settles over an area and does not move for several days.

Air mass

Cherry trees blooming during New York City's heat wave in January 2007.

Winter Heat Waves

Temperatures during winter heat waves can get surprisingly high. In January 2003, in Los Angeles, California, the temperatures, which are normally around 68°F (20°C), rose well over 90°F (32°C). In January 2007, cherry trees in New York City started to bloom three months early because of a winter heat wave. Temperatures hit 72°F (22°C).

Children try to stay cool during a heat wave in Pakistan by covering their heads with wet towels.

Deadly Heat

High heat can put stress on the body. Heat exhaustion can happen to people who spend a lot of time outside during hot weather. They start sweating a lot, become dizzy, and may faint. Heatstroke is even more dangerous. This is when a person's body temperature is more than 104°F (40°C) and their heart beats really fast. A person with heatstroke needs quick medical attention.

Broiling Buildings and Ruined Roads

During heat waves, homes and buildings can become extra hot. People often crank up air conditioners, which use lots of electricity. This can cause power failures called blackouts. Heat waves also affect bridges and roads. Metal and concrete swell up when they get hot. As the materials push apart, they crack and break. But one of the biggest problems with heat waves is that they can also contribute to droughts.

Summer heat waves can often cause roads to break and buckle.

In addition to heat waves, cutting too many trees can also contribute to droughts.

Extreme heat dried up most of the water in this reservoir in Spain.

When Droughts Happen

Different places on Earth get different amounts of rain. The more precipitation a place gets, the more plants and animals are able to make it their home. However, if drought hits an area and lasts long enough, it can cause serious problems for plants and animals living there, and even for people. Without a regular supply of fresh water, most living things can die.

Why Do Droughts Happen?

Droughts happen when an area is drier than usual because it gets too little rain or snow. Droughts are more common in the summer, but they can happen any time of the year. Like heat waves, droughts are caused by a high-pressure air mass that stays over an area. This blocks new air masses that would normally bring in moisture.

Meteorologists use a rain gauge to measure rainfall. It helps them keep track of any changes in normal amounts of rainfall.

Ring Around the Drought

Take a close look at a tree stump, and you'll see a set of rings that starts small at the center and gets larger. Each year a tree grows a new ring. During wet years, trees grow faster so the rings are thicker. During years with droughts, trees grow slower so the rings are thinner. By studying the thickness of tree rings, scientists can tell when droughts happened in the past.

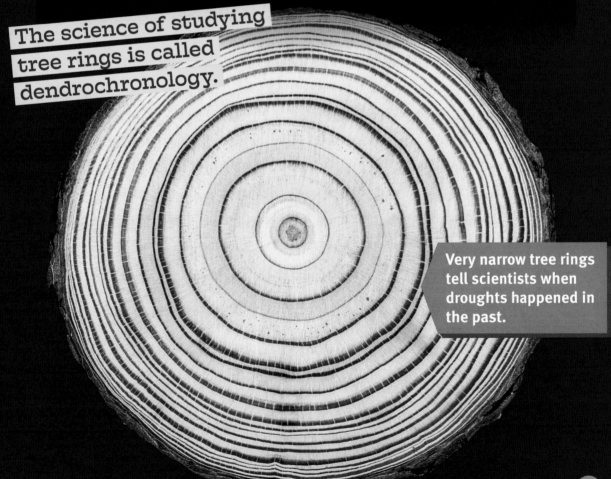

The science of studying tree rings is called dendrochronology.

Very narrow tree rings tell scientists when droughts happened in the past.

During the Dust Bowl in the 1930s, large dust storms blew across the midwestern United States when a severe drought hit.

The Impacts of Droughts

Droughts cause many problems for people and the environment. The longer they last, the worse the problems become. When severe droughts happen, streams, lakes, and **reservoirs** dry up. When reservoirs are dry, there is less water for people to drink. On a farm, crops and animals die if there's no water. When plants die, the roots that hold soil in place are destroyed, so strong winds can blow the soil away.

Wildfires

Droughts are dangerous for another reason: They often lead to wildfires. During droughts, grasses and leaves on forest floors dry out and can easily catch fire. All that is needed is a spark or a lightning strike. Wildfires spread quickly and are difficult to control. They are especially difficult to fight during summer heat waves when the air is hot and dry. In recent years, hundreds of deadly wildfires have happened following severe droughts.

A wildfire burns out of control as firefighters try to fight the flames.

Droughts in the Rain Forest

When a rain forest has a drought, trees lose leaves or even die. This means more sunlight gets into the lower parts of the forest, making it drier than it should be. A dry forest makes it hard for plants and animals to survive. The Amazon rain forest in South America and the Tongass National Forest in Alaska are two rain forests that have experienced droughts.

The Amazon Rain Forest

The Amazon rain forest is the largest rain forest on Earth. It is usually one of the wettest places on the planet. But over the last 30 years, there have been several major droughts in the Amazon, and they have been happening more frequently. Many scientists believe that people are to blame. Trees put a lot of water vapor into the air. Unfortunately, people have been cutting down trees in huge numbers in the Amazon. Fewer trees means less water vapor in the air, which leads to less rain.

This area of the Amazon has been badly affected by drought. There are few trees, and the land is very dry.

Tongass National Forest

The Tongass National Forest is the largest temperate rain forest in the world. Since 2018, it has experienced droughts—and some have been extreme. The droughts hurt salmon habitats, destroy trees, and cause water supply problems for people who live in the area.

Droughts have caused low water levels at Purple Lake in Tongass National Forest.

It can feel about 15 degrees cooler
in the shade than in the sun.

Keeping Your Cool

During a heat wave, it's important to stay cool and avoid going out during the warmest part of the day—usually in the mid to late afternoon. There are many things that you can do to beat the heat and keep yourself and your family and friends safe, from turning on the air-conditioning to drinking lots of water!

A fan can keep you cool on a hot summer day.

A playground is shaded to help kids stay out of the sun.

Staying In and Slowing Down

If you can, stay indoors in an air-conditioned place when it's extremely hot. If your home is not air-conditioned, go to a library or other public place that is cool. During heat waves, many communities open cooling centers for people to use. If you must go outside, find a shady spot or use an umbrella to block the sun. Then, take it slow. Why? Physical activity makes you hotter. Do activities that don't require much movement.

Dressing Light

Staying out of the sun is not the only way to stay cool. The way you dress also makes a difference. Wearing a hat protects your head and face from the sun, and loose-fitting shorts and T-shirts allow air to flow over your skin. Wearing lighter colors helps, too. Why? Dark colors, such as navy blue and black, soak up more of the sun's heat than light colors do.

Hats, sunglasses, and light clothes or special sun-blocking T-shirts are perfect for a day at the beach.

The Wonders of Water

One quick way to cool off during a heat wave is to get wet! Hit the beach, take a dip in a pool, or run under a sprinkler. When moisture evaporates off your skin, it lowers your body temperature. This is one reason we sweat. Sweating cools your skin, but it also causes your body to lose a lot of water. When it's hot, drink lots of fluids, including water, so you don't get **dehydrated**!

Volunteers hand out cold juice on a very hot summer day in Pakistan.

Watering cans can be used for both indoor and outdoor plants.

Turn Off the Tap

Saving water is the most important thing to do during a heat wave or drought. The local news stations often report on rules to follow until a heat wave or drought is over: Take short showers instead of baths, don't let the water run while washing hands or brushing teeth, and don't wash cars or walkways. It's also best to water flowers with a watering can instead of letting a hose run.

Since 1950 the number of heat waves and severe droughts in the world has steadily gone up.

Climate scientists tell us that we should be prepared for more record-breaking temperatures in the future.

What the Future Holds

Climate scientists collect rainfall and temperature data from around the world. This data is used to predict what the climate will be like in the future using computer models. Many of these models show that heat waves and droughts will get worse in the next 100 years because of climate change. But there are some actions we can take to help change this dry, hot forecast.

Dealing with Droughts

Scientists are working on new technologies to lessen the frequency of droughts and help people deal with them when they do happen. New watering systems have been developed so farmers can use less water while growing the same amount of crops. Countries like Saudi Arabia and Israel are using special technologies to remove salt from seawater to make it fresh so that people can use it.

Timeline of Heat Waves Throughout History

AUGUST 1896
Ten days of extreme heat and crowded streets and buildings lead to dangerous living conditions in New York City.

SUMMER 1936
Higher than normal temperatures and a long drought spread across the midwestern United States and central Canada.

JULY 1995
A major heat wave in Chicago, Illinois, lasts for five days and causes temperatures to go over 100°F (38°C).

Slowing the Heat

Carbon dioxide is one of the gases in the atmosphere that increases the greenhouse effect. This leads to higher temperatures on Earth. Power plants that burn fossil fuels to make electricity release extra carbon dioxide into the air. Greener energy sources such as solar panels and wind turbines do not release carbon dioxide. Replacing power plants with greener sources of electricity can lower the amount of carbon dioxide that is released and decrease the effect of global warming.

SUMMER 2003
Record-breaking temperatures in Europe wipe out farms and kill livestock.

SUMMER 2010
In Russia, temperatures soar over 100°F (38°C), leading to the deaths of more than 50,000 people.

JUNE 2015
Temperatures reaching 120°F (49°C) are hot enough to melt pavement in India and Pakistan.

SUMMER 2020
Heat waves from June to August are responsible for more than 2,500 deaths in Great Britain.

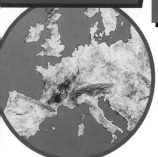

The Power of Trees

Trees are important when it comes to fighting the effects of heat waves and droughts. Tree roots pull water from the soil and release it as water vapor in the air. More water vapor means that there is a better chance for rain or snow. When trees grow, they take carbon dioxide out of the air. Many scientists believe that planting millions of trees will help reduce greenhouse gases in the atmosphere, leading to a cooler planet.

Volunteers plant a tree in a city park. Trees provide shade and can improve air quality.

Our Role in the Future

Most climate scientists agree that it's time to take action to keep our planet safe from heat waves, droughts, and other severe weather. Many people, including kids, are already making a difference not only by saving water, but also by saving electricity, recycling, and speaking up for the environment. They want to save the planet they will one day be in charge of. How about you?

Mapping Droughts

Scientists collect lots of data about rainfall and droughts. Using this data, they create maps that show where droughts are happening and how severe they are. This map shows where droughts occurred in the United States on September 29, 2020. The colors on the map match the key to show how harsh each drought was. Use the map and key to answer questions on the following page.

Drought Conditions in the United States

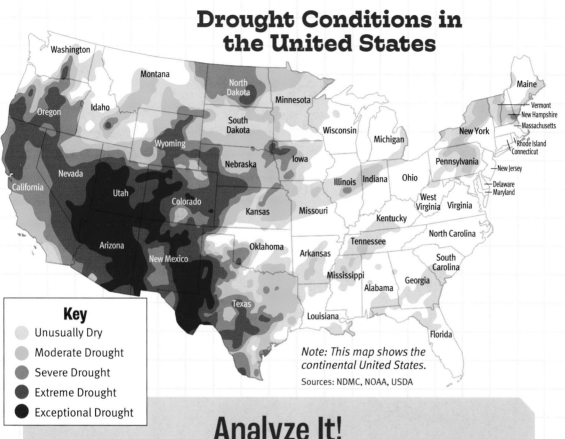

Washington
Montana
North Dakota
Minnesota
Maine
Oregon
Idaho
South Dakota
Wisconsin
Michigan
New York
Vermont
New Hampshire
Massachusetts
Wyoming
Iowa
Pennsylvania
Rhode Island
Connecticut
Nevada
Nebraska
Illinois
Indiana
Ohio
New Jersey
California
Utah
Colorado
Kansas
Missouri
West Virginia
Virginia
Delaware
Maryland
Kentucky
Arizona
New Mexico
Oklahoma
Arkansas
Tennessee
North Carolina
South Carolina
Texas
Mississippi
Alabama
Georgia
Louisiana
Florida

Key
- Unusually Dry
- Moderate Drought
- Severe Drought
- Extreme Drought
- Exceptional Drought

Note: This map shows the continental United States.
Sources: NDMC, NOAA, USDA

Analyze It!

1. What part of the country had the worst drought conditions?

2. What part of the country was least affected by drought?

3. What were the drought conditions in Vermont?

4. What were the drought conditions in Illinois?

5. In what part of the country do you think it would be more difficult to start a new farm?

ANSWERS: 1. Southwestern United States. 2. Mid-Atlantic and southeastern United States 3. Unusually dry and moderate 4. Unusually dry, moderate, and severe 5. The Southwest because there are severe droughts and therefore not enough water for crops.

Measuring Evaporation

During the summer, droughts cause water to evaporate from reservoirs. This leaves less water for people to use. Is there a way to reduce the amount of evaporation to save water?

Materials

Measuring cup

Water

Two bowls that are the same size and material

Sheet of paper

Pen or pencil

Directions

1 Use the measuring cup to fill each bowl with ½ cup of water.

2 Place the two bowls side by side on a sunny counter or windowsill.

3 Place the sheet of paper over one of the bowls, making sure that it covers the entire bowl but does not touch the water.

4 Allow the two bowls to sit for at least one full day. Make sure they are left alone.

5 Pour the water from the uncovered bowl back into the measuring cup. Record how much water is left. Empty the measuring cup and repeat with the covered bowl of water. Which bowl had more water in it?

Explain It!

Based on this activity and what you learned in this book, can you explain the results of this experiment?

True Statistics

Highest temperature recorded on Earth: 134°F (56.7°C) on July 10, 1913, in Death Valley, California

Highest temperature recorded in Siberia, which is normally cold year-round: 100.4°F (38°C) on June 20, 2020, in the town of Verkhoyansk

Highest temperature in Antarctica, the coldest place on Earth: 64.9°F (18.3°C) in February 2020

Warmest month in the United States: July 1936. Fourteen states set record-high temperatures over 109°F (42.8°C).

Driest place on Earth where people live: Arica, Chile. It gets less than 0.03 inch (0.08 cm) of rain and snow per year.

Most severe drought in North America: Three different drought periods in 1934, 1936, and from 1939 to 1940 destroyed over 100 million acres of farmland and forced thousands of people to leave their homes.

Did you find the truth?

(T) Sometimes human activity can cause droughts to happen.

(F) Heat waves only happen in the summer.

Resources

Other books in this series:

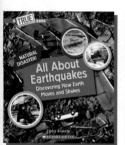   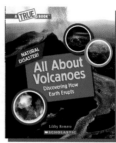 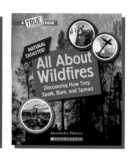

You can also look at:

Challen, Paul. *Drought and Heat Wave Alert!* St. Catharines, Ontario: Crabtree Publishing, 2004.

Grunbaum, Mara. *The Greenhouse Effect*. New York: Children's Press, 2020.

McDaniel, Melissa. *Understanding Climate Change: Facing a Warming World*. New York: Children's Press, 2020.

Meister, Cari. *Disaster Zone: Droughts*. Minneapolis, Minnesota: Jump!, 2016.

Rajczak, Michael. *Deadly Droughts: Where's the Water?* New York: Gareth Stevens Publishing, 2016.

Wendorff, Anne. *Extreme Weather: Droughts*. Minneapolis, Minnesota: Bellwether Media, 2016.

Glossary

atmosphere (AT-muhs-feer) the mixture of gases that surrounds a planet

climate (KLYE-mit) the usual weather that happens in a place

climate regions (KLYE-mit REE-juhns) places that have a similar climate

condensation (kahn-den-SAY-shuhn) the act of a gas turning into a liquid

dehydrated (dee-HYE-dray-tid) when something has the water removed from it

evaporation (i-vap-uh-RAY-shuhn) the act of changing from a liquid to a gas

greenhouse effect (GREEN-hous i-FEKT) the warming of the atmosphere caused by gases such as carbon dioxide that prevent some heat from leaving

humid (HYOO-mid) weather that is moist and usually very warm, in a way that is uncomfortable

meteorologists (mee-tee-uh-RAH-luh-jists) scientists who study the weather

precipitation (pri-sip-i-TAY-shuhn) rain, hail, or snow

reservoirs (REZ-ur-vwahrz) human-made lakes used to store water for people to use

water vapor (WAH-tur VAY-pur) water in the form of a gas

Index

Page numbers in **bold** indicate illustrations.

About the Author

Steve Tomecek is an award-winning author of more than 50 nonfiction books dealing with science and technology for kids and teachers. He graduated from Queens College and began his career as a geologist working as an environmental planner and conducting research on climate change. For the past 30 years he has served as the executive director of Science Plus, Inc., a company based in New York City that provides science programs and consulting services for schools and organizations throughout the United States. Visit his website at www.dirtmeister.com.